Editor
Gisela Lee

Managing Editor
Karen Goldfluss, M.S. Ed.

Editor-in-Chief
Sharon Coan, M.S. Ed.

Art Director
CJae Froshay

Art Coordinator
Kevin Barnes

Cover Artist
Barb Lorseyedi

Imaging
Alfred Lau
James Edward Grace
Rosa C. See

Product Manager
Phil Garcia

Publishers
Rachelle Cracchiolo, M.S. Ed.
Mary Dupuy Smith, M.S. Ed.

Practice Makes Perfect

Division

GRADE 3

D0339775

Author

Robert Smith

Teacher Created Materials, Inc.
6421 Industry Way
Westminster, CA 92683
www.teachercreated.com

ISBN-0-7439-3323-X

©2002 Teacher Created Materials, Inc.
Made in U.S.A.

Table of Contents

Introduction

The old adage "practice makes perfect" can really hold true for your child and his or her education. The more practice and exposure your child has with concepts being taught in school, the more success he or she is likely to find. For many parents, knowing how to help your children can be frustrating because the resources may not be readily available. As a parent it is also difficult to know where to focus your efforts so that the extra practice their child receives at home supports what he or she is learning in school.

This book has been designed to help parents and teachers reinforce basic skills with their children. *Practice Makes Perfect* reviews basic math skills for children in the third grade. The math focus is on division. While it would be impossible to include all concepts taught in the third grade in this book, the following basic objectives are reinforced through practice exercises. These objectives support math standards established on a district, state, or national level. (Refer to the Table of Contents for the specific objectives of each practice page.)

- division facts to 20
- division tables
- word problems
- rules of divisibility

- dividing with money
- one-digit division with and without remainders
- two-digit division with and without remainders

There are 36 practice pages organized sequentially, so children can build their knowledge from more basic skills to higher-level math skills. To correct the practice pages in this book, use the answer key provided on pages 47 and 48. Six practice tests follow the practice pages. These provide children with multiple-choice test items to help prepare them for standardized tests administered in schools. As children complete a problem, they fill in the correct letter among the answer choices. An optional "bubble-in" answer sheet has also been provided on page 46. This answer sheet is similar to those found on standardized tests. As your child completes each test, he or she can fill in the correct bubbles on the answer sheet.

How to Make the Most of This Book

Here are some useful ideas for optimizing the practice pages in this book:

- Set aside a specific place in your home to work on the practice pages. Keep it neat and tidy with materials on hand.

- Set up a certain time of day to work on the practice pages. This will establish consistency. An alternative is to look for times in your day or week that are less hectic and are more conducive to practicing skills.

- Keep all practice sessions with your child positive and constructive. If the mood becomes tense or you and your child are frustrated, set the book aside and look for another time to practice with your child.

- Help with instructions if necessary. If your child is having difficulty understanding what to do or how to get started, work through the first problem with him or her.

- Review the work your child has done. This serves as reinforcement and provides further practice.

- Allow your child to use whatever writing instruments he or she prefers. For example, colored pencils can add variety and pleasure to drill work.

- Pay attention to the areas in which your child has the most difficulty. Provide extra guidance and exercises in those areas. Allowing children to use drawings and manipulatives, such as coins, tiles, game markers, or flash cards can help them grasp difficult concepts more easily.

- Look for ways to make real-life applications to the skills being reinforced.

Practice 1

The multiplication chart shown here can be used to find any basic multiplication or division fact until you have learned them all.

One of the best ways to learn the facts is to practice using the chart.

Columns

Rows	1	2	3	4	5	6	7	8	9	10	11	12
1	1	2	3	4	5	6	7	8	9	10	11	12
2	2	4	6	8	10	12	14	16	18	20	22	24
3	3	6	9	12	15	18	21	24	27	30	33	36
4	4	8	12	16	20	24	28	32	36	40	44	48
5	5	10	15	20	25	30	35	40	45	50	55	60
6	6	12	18	24	30	36	42	48	54	60	66	72
7	7	14	21	28	35	42	49	56	63	70	77	84
8	8	16	24	32	40	48	56	64	72	80	88	96
9	9	18	27	36	45	54	63	72	81	90	99	108
10	10	20	30	40	50	60	70	80	90	100	110	120
11	11	22	33	44	55	66	77	88	99	110	121	132
12	12	24	36	48	60	72	84	96	108	120	132	144

Read **across** for the **Rows**.

Read **up** or **down** for the **Columns**.

Note: To find how many times 7 divides into 63, run one finger across the 7 row until you come to 63 and run a finger up the column with 63 until you come to the top number which is 9. The answer is that 7 divides into 63 exactly 9 times.

Directions: Use the rows on the multiplication chart to help you find the missing numbers. (Go backwards.)

1.   (36, 33, 30, 27, 24, _____ , _____ , _____ , _____ , _____ , _____ , _____)

2.   (24, 22, 20, 18, 16, 14, _____ , _____ , _____ , _____ , _____ , _____)

3.   (48, 44, 40, 36, _____ , _____ , _____ , _____ , _____ , _____ , _____ , _____)

4.   (60, 55, 50, 45, 40, _____ , _____ , _____ , _____ , _____ , _____ , _____)

5.   (49, 42, 35, 28, _____ , _____ , _____)

6.   (48, 40, 32, 24, _____ , _____)

7.   (120, 108, 96, 84, 72, 60, _____ , _____ , _____ , _____)

8.   Which row has a zero in every number? _____

Practice 2 ༄ ༄ ༄ ༄ ༄ ༄ ༄ ༄ ༄ ༄ ༄ ༄ ༄

Directions: Use the columns on the multiplication/division chart to help you find the missing numbers. (Go backwards.)

1.	2.	3.	4.	5.
48	84	60	120	12
44	77	55	110	11
40	70	50	100	10
36	63	45	90	9
32	56	40	80	8
——	——	——	——	——
——	——	——	——	——
——	——	——	——	——
——	——	——	——	——
——	——	——	——	——
——	——	——	——	——

6. Which column has a zero in every number? _____

Directions: Use the multiplication/division chart to help you find the answers to these problems. The first two are done for you.

7. How many 8's can you subtract from 40?

   $48 - 8 - 8 - 8 - 8 - 8 - 8 = 0$

   You can subtract six 8's from 48.

8. How many 5's can you subtract from 20?

   $20 - 5 - 5 - 5 - 5 = 0$

   You can subtract four 5's from 20.

9. How many 6's can you subtract from 36?

   $36 - 6 - 6 - 6 - 6 - 6 - 6 = 0$

   _____

10. How many 7's can you subtract from 28?

    $28 - $

    _____

11. How many 3's can you subtract from 15?

    $15 - $

    _____

12. How many 4's can you subtract from 24?

    $24 - $

    _____

Practice 3 ᓚ ᓗ ᓚ ᓗ ᓚ ᓗ ᓚ ᓗ ᓚ ᓗ ᓚ ᓗ ᓚ

Directions: Do these problems. Use your multiplication/division chart if you are unsure of your division facts.

1. $1\overline{)5}$ 2. $2\overline{)10}$ 3. $5\overline{)5}$ 4. $1\overline{)4}$

5. $5\overline{)5}$ 6. $2\overline{)8}$ 7. $5\overline{)15}$ 8. $2\overline{)6}$

9. $2\overline{)12}$ 10. $5\overline{)20}$ 11. $1\overline{)3}$ 12. $5\overline{)25}$

13. $5\overline{)35}$ 14. $2\overline{)14}$ 15. $2\overline{)8}$ 16. $1\overline{)9}$

17. $1\overline{)7}$ 18. $5\overline{)40}$ 19. $1\overline{)4}$ 20. $2\overline{)20}$

21. $5\overline{)45}$ 22. $1\overline{)7}$ 23. $2\overline{)16}$ 24. $5\overline{)50}$

25. $1\overline{)8}$ 26. $2\overline{)18}$ 27. $1\overline{)10}$ 28. $5\overline{)30}$

29. $5\overline{)10}$ 30. $1\overline{)9}$ 31. $2\overline{)20}$ 32. $2\overline{)4}$

33. $1\overline{)5}$ 34. $2\overline{)8}$ 35. $5\overline{)20}$ 36. $1\overline{)9}$

37. $2\overline{)10}$ 38. $5\overline{)30}$ 39. $1\overline{)11}$ 40. $2\overline{)16}$

Practice 4 ꙮ ꙮ ꙮ ꙮ ꙮ ꙮ ꙮ ꙮ ꙮ ꙮ ꙮ ꙮ ꙮ

Directions: Do these problems. Use your multiplication/division chart if you are unsure of your division facts.

1. $3\overline{)6}$ 2. $4\overline{)8}$ 3. $6\overline{)12}$ 4. $3\overline{)12}$

5. $4\overline{)12}$ 6. $6\overline{)18}$ 7. $6\overline{)24}$ 8. $4\overline{)36}$

9. $3\overline{)15}$ 10. $3\overline{)21}$ 11. $6\overline{)72}$ 12. $6\overline{)48}$

13. $4\overline{)32}$ 14. $4\overline{)24}$ 15. $3\overline{)27}$ 16. $3\overline{)36}$

17. $6\overline{)54}$ 18. $6\overline{)42}$ 19. $4\overline{)4}$ 20. $4\overline{)20}$

21. $3\overline{)30}$ 22. $3\overline{)3}$ 23. $3\overline{)12}$ 24. $4\overline{)36}$

25. $4\overline{)8}$ 26. $6\overline{)18}$ 27. $4\overline{)16}$ 28. $6\overline{)30}$

29. $4\overline{)28}$ 30. $4\overline{)44}$ 31. $4\overline{)48}$ 32. $4\overline{)40}$

33. $3\overline{)9}$ 34. $3\overline{)36}$ 35. $3\overline{)24}$ 36. $6\overline{)36}$

37. $6\overline{)54}$ 38. $6\overline{)60}$ 39. $6\overline{)66}$ 40. $6\overline{)72}$

Practice 5 ᵔ ᵔ ᵔ ᵔ ᵔ ᵔ ᵔ ᵔ ᵔ ᵔ ᵔ ᵔ ᵔ ᵔ ᵔ

Directions: Do these problems. Use your multiplication/division chart if you are unsure of your division facts.

1. $7\overline{)14}$ 　　2. $7\overline{)21}$ 　　3. $8\overline{)16}$ 　　4. $9\overline{)18}$

5. $9\overline{)27}$ 　　6. $9\overline{)45}$ 　　7. $8\overline{)24}$ 　　8. $9\overline{)36}$

9. $7\overline{)35}$ 　　10. $7\overline{)28}$ 　　11. $7\overline{)63}$ 　　12. $8\overline{)48}$

13. $8\overline{)32}$ 　　14. $9\overline{)54}$ 　　15. $9\overline{)72}$ 　　16. $8\overline{)64}$

17. $9\overline{)63}$ 　　18. $7\overline{)42}$ 　　19. $7\overline{)7}$ 　　20. $8\overline{)56}$

21. $7\overline{)49}$ 　　22. $7\overline{)56}$ 　　23. $8\overline{)56}$ 　　24. $8\overline{)96}$

25. $7\overline{)63}$ 　　26. $9\overline{)63}$ 　　27. $8\overline{)64}$ 　　28. $8\overline{)40}$

29. $7\overline{)70}$ 　　30. $7\overline{)77}$ 　　31. $7\overline{)84}$ 　　32. $8\overline{)40}$

33. $8\overline{)8}$ 　　34. $9\overline{)36}$ 　　35. $9\overline{)63}$ 　　36. $9\overline{)36}$

37. $9\overline{)54}$ 　　38. $9\overline{)72}$ 　　39. $9\overline{)81}$ 　　40. $8\overline{)72}$

#3323 Practice Makes Perfect: Division　　　　　*© Teacher Created Materials, Inc.*

Practice 6 ⟳ ⟳ ⟳ ⟳ ⟳ ⟳ ⟳ ⟳ ⟳ ⟳ ⟳ ⟳ ⟳

Directions: Do these problems. Use your multiplication/division chart if you are unsure of your division facts.

1. $5\overline{)5}$

2. $5\overline{)10}$

3. $4\overline{)8}$

4. $4\overline{)4}$

5. $5\overline{)15}$

6. $2\overline{)10}$

7. $5\overline{)25}$

8. $2\overline{)18}$

9. $3\overline{)12}$

10. $5\overline{)35}$

11. $1\overline{)9}$

12. $5\overline{)45}$

13. $4\overline{)32}$

14. $2\overline{)16}$

15. $2\overline{)6}$

16. $1\overline{)7}$

17. $1\overline{)4}$

18. $5\overline{)60}$

19. $4\overline{)40}$

20. $2\overline{)22}$

21. $5\overline{)55}$

22. $1\overline{)6}$

23. $2\overline{)16}$

24. $5\overline{)20}$

25. $2\overline{)8}$

26. $2\overline{)12}$

27. $5\overline{)40}$

28. $4\overline{)32}$

29. $3\overline{)15}$

30. $3\overline{)9}$

31. $4\overline{)20}$

32. $4\overline{)44}$

33. $3\overline{)6}$

34. $3\overline{)24}$

35. $4\overline{)28}$

36. $4\overline{)36}$

37. $5\overline{)50}$

38. $4\overline{)36}$

39. $4\overline{)48}$

40. $4\overline{)16}$

Practice 7 ๑ ๑ ๑ ๑ ๑ ๑ ๑ ๑ ๑ ๑ ๑ ๑ ๑

Directions: Do these problems. Use your multiplication/division chart if you are unsure of your division facts.

1. $9\overline{)18}$ 2. $7\overline{)28}$ 3. $6\overline{)18}$ 4. $7\overline{)35}$

5. $9\overline{)36}$ 6. $9\overline{)54}$ 7. $6\overline{)36}$ 8. $9\overline{)63}$

9. $8\overline{)32}$ 10. $7\overline{)14}$ 11. $7\overline{)49}$ 12. $6\overline{)48}$

13. $8\overline{)48}$ 14. $6\overline{)54}$ 15. $8\overline{)80}$ 16. $8\overline{)88}$

17. $7\overline{)63}$ 18. $7\overline{)49}$ 19. $9\overline{)81}$ 20. $7\overline{)56}$

21. $8\overline{)64}$ 22. $9\overline{)90}$ 23. $8\overline{)96}$ 24. $9\overline{)99}$

25. $6\overline{)12}$ 26. $6\overline{)36}$ 27. $6\overline{)72}$ 28. $8\overline{)48}$

29. $8\overline{)96}$ 30. $9\overline{)108}$ 31. $8\overline{)40}$ 32. $6\overline{)36}$

33. $8\overline{)24}$ 34. $6\overline{)24}$ 35. $6\overline{)66}$ 36. $9\overline{)63}$

37. $9\overline{)45}$ 38. $9\overline{)54}$ 39. $8\overline{)80}$ 40. $9\overline{)72}$

Practice 8 ꙮ ꙮ ꙮ ꙮ ꙮ ꙮ ꙮ ꙮ ꙮ ꙮ ꙮ ꙮ ꙮ

Directions: Fill in the missing factors. Use your multiplication/division chart, if you are unsure of your facts.

1. $10\overline{)20}$
2. $11\overline{)55}$
3. $12\overline{)36}$
4. $10\overline{)30}$

5. $11\overline{)44}$
6. $10\overline{)60}$
7. $11\overline{)33}$
8. $10\overline{)90}$

9. $11\overline{)55}$
10. $12\overline{)72}$
11. $11\overline{)66}$
12. $10\overline{)100}$

13. $10\overline{)70}$
14. $12\overline{)48}$
15. $11\overline{)22}$
16. $10\overline{)110}$

17. $12\overline{)60}$
18. $12\overline{)36}$
19. $11\overline{)88}$
20. $11\overline{)110}$

21. $10\overline{)20}$
22. $10\overline{)40}$
23. $12\overline{)96}$
24. $11\overline{)99}$

25. $11\overline{)66}$
26. $12\overline{)36}$
27. $12\overline{)84}$
28. $12\overline{)48}$

29. $12\overline{)96}$
30. $11\overline{)121}$
31. $10\overline{)40}$
32. $12\overline{)84}$

33. $12\overline{)12}$
34. $11\overline{)44}$
35. $12\overline{)60}$
36. $12\overline{)144}$

37. $10\overline{)50}$
38. $11\overline{)77}$
39. $11\overline{)88}$
40. $10\overline{)120}$

Practice 9

Directions: Do these problems. Use your multiplication/division chart if you are unsure of your division facts.

1. $4\overline{)16}$ 2. $6\overline{)30}$ 3. $10\overline{)20}$ 4. $6\overline{)36}$

5. $12\overline{)36}$ 6. $7\overline{)49}$ 7. $7\overline{)56}$ 8. $7\overline{)21}$

9. $5\overline{)30}$ 10. $5\overline{)55}$ 11. $8\overline{)56}$ 12. $6\overline{)66}$

13. $4\overline{)48}$ 14. $9\overline{)54}$ 15. $4\overline{)24}$ 16. $6\overline{)30}$

17. $9\overline{)54}$ 18. $7\overline{)63}$ 19. $9\overline{)108}$ 20. $7\overline{)35}$

21. $7\overline{)28}$ 22. $4\overline{)32}$ 23. $3\overline{)15}$ 24. $5\overline{)15}$

25. $3\overline{)24}$ 26. $4\overline{)24}$ 27. $8\overline{)24}$ 28. $12\overline{)24}$

29. $9\overline{)90}$ 30. $3\overline{)27}$ 31. $4\overline{)40}$ 32. $3\overline{)36}$

33. $11\overline{)33}$ 34. $6\overline{)54}$ 35. $9\overline{)63}$ 36. $8\overline{)64}$

37. $5\overline{)45}$ 38. $5\overline{)40}$ 39. $5\overline{)25}$ 40. $3\overline{)27}$

Practice 10 ♪ ☾ ♪ ☾ ♪ ☾ ♪ ☾ ♪ ☾ ♪ ☾ ♪ ☾ ♪ ☾

Directions: Do these problems. Use your multiplication/division chart if you are unsure of your division facts.

1. $1\overline{)10}$ 2. $6\overline{)48}$ 3. $10\overline{)90}$ 4. $4\overline{)36}$

5. $12\overline{)60}$ 6. $3\overline{)27}$ 7. $8\overline{)56}$ 8. $7\overline{)49}$

9. $5\overline{)55}$ 10. $10\overline{)40}$ 11. $12\overline{)84}$ 12. $11\overline{)99}$

13. $4\overline{)24}$ 14. $3\overline{)33}$ 15. $4\overline{)36}$ 16. $6\overline{)36}$

17. $7\overline{)63}$ 18. $9\overline{)63}$ 19. $9\overline{)108}$ 20. $12\overline{)108}$

21. $3\overline{)18}$ 22. $2\overline{)18}$ 23. $5\overline{)15}$ 24. $3\overline{)27}$

25. $4\overline{)36}$ 26. $6\overline{)36}$ 27. $9\overline{)36}$ 28. $12\overline{)36}$

29. $9\overline{)72}$ 30. $12\overline{)72}$ 31. $8\overline{)72}$ 32. $6\overline{)72}$

33. $7\overline{)63}$ 34. $9\overline{)18}$ 35. $9\overline{)63}$ 36. $8\overline{)32}$

37. $8\overline{)48}$ 38. $6\overline{)48}$ 39. $12\overline{)48}$ 40. $4\overline{)48}$

Practice 11 ⟳ ⟳ ⟳ ⟳ ⟳ ⟳ ⟳ ⟳ ⟳ ⟳ ⟳ ⟳ ⟳ ⟳ ⟳

Directions: Fill in the missing factors. Use your multiplication/division chart, if you are unsure of your facts.

1. $8 \times \underline{} = 48$

2. $6 \times \underline{} = 48$

3. $12 \times \underline{} = 48$

4. $9 \times \underline{} = 63$

5. $\underline{} \times 7 = 63$

6. $7 \times \underline{} = 63$

7. $4 \times \underline{} = 36$

8. $9 \times \underline{} = 36$

9. $\underline{} \times 6 = 36$

10. $\underline{} \times 7 = 42$

11. $\underline{} \times 6 = 42$

12. $6 \times \underline{} = 42$

13. $7 \times \underline{} = 56$

14. $\underline{} \times 8 = 56$

15. $\underline{} \times 7 = 56$

16. $\underline{} \times 6 = 54$

17. $9 \times \underline{} = 54$

18. $6 \times \underline{} = 54$

19. $48 \div 6 = \underline{}$

20. $48 \div 8 = \underline{}$

21. $48 \div 12 = \underline{}$

22. $64 \div 8 = \underline{}$

23. $81 \div 9 = \underline{}$

24. $36 \div 12 = \underline{}$

25. $49 \div 7 = \underline{}$

26. $35 \div 7 = \underline{}$

27. $84 \div 12 = \underline{}$

28. $36 \div \underline{} = 3$

29. $36 \div \underline{} = 6$

30. $84 \div \underline{} = 12$

31. $88 \div \underline{} = 8$

32. $144 \div \underline{} = 12$

33. $66 \div \underline{} = 6$

34. $54 \div \underline{} = 6$

35. $54 \div \underline{} = 9$

36. $32 \div \underline{} = 8$

Practice 12 ꙮ ꙮ ꙮ ꙮ ꙮ ꙮ ꙮ ꙮ ꙮ ꙮ ꙮ ꙮ ꙮ ꙮ

Directions: Do these problems. Use your multiplication/division chart if you are unsure of your division facts. The first two are done for you.

1. $\begin{array}{r} 5 \text{ R1} \\ 3\overline{)16} \\ -15 \\ \hline 1 \end{array}$

2. $\begin{array}{r} 4 \text{ R2} \\ 4\overline{)18} \\ -16 \\ \hline 2 \end{array}$

3. $5\overline{)21}$

4. $3\overline{)19}$

5. $4\overline{)37}$

6. $7\overline{)36}$

7. $5\overline{)36}$

8. $9\overline{)37}$

9. $4\overline{)35}$

10. $6\overline{)43}$

11. $7\overline{)50}$

12. $4\overline{)19}$

13. $6\overline{)49}$

14. $7\overline{)50}$

15. $4\overline{)39}$

16. $3\overline{)22}$

17. $6\overline{)55}$

18. $6\overline{)38}$

19. $4\overline{)34}$

20. $5\overline{)19}$

21. $8\overline{)65}$

22. $7\overline{)30}$

23. $7\overline{)31}$

24. $3\overline{)13}$

Practice 13

Directions: Do these problems. Use your multiplication/division chart if you are unsure of your division facts. The first two are done for you.

1. $\begin{array}{r} 6\ \text{R1} \\ 4\overline{)25} \\ -24 \\ \hline 1 \end{array}$

2. $\begin{array}{r} 9\ \text{R3} \\ 6\overline{)57} \\ -54 \\ \hline 3 \end{array}$

3. $2\overline{)17}$

4. $4\overline{)19}$

5. $5\overline{)42}$

6. $5\overline{)26}$

7. $3\overline{)23}$

8. $6\overline{)19}$

9. $5\overline{)39}$

10. $7\overline{)48}$

11. $6\overline{)51}$

12. $5\overline{)29}$

13. $8\overline{)42}$

14. $6\overline{)38}$

15. $9\overline{)46}$

16. $9\overline{)44}$

17. $7\overline{)64}$

18. $7\overline{)62}$

19. $8\overline{)34}$

20. $7\overline{)19}$

21. $6\overline{)28}$

22. $4\overline{)31}$

23. $5\overline{)12}$

24. $8\overline{)17}$

 #3323 *Practice Makes Perfect: Division*

Practice 14 ౨ ౿ ౨ ౿ ౨ ౿ ౨ ౿ ౨ ౿ ౨ ౿ ౨ ౨ ౿

Directions: Do these problems. Use your multiplication/division chart if you are unsure of your division facts. The first two are done for you.

1. $\begin{array}{r} 3 \text{ R2} \\ 9\overline{)29} \\ -27 \\ \hline 2 \end{array}$

2. $\begin{array}{r} 8 \text{ R3} \\ 8\overline{)67} \\ -64 \\ \hline 3 \end{array}$

3. $3\overline{)28}$

4. $5\overline{)46}$

5. $7\overline{)51}$

6. $8\overline{)43}$

7. $6\overline{)23}$

8. $5\overline{)29}$

9. $7\overline{)47}$

10. $8\overline{)18}$

11. $9\overline{)31}$

12. $10\overline{)36}$

13. $11\overline{)47}$

14. $7\overline{)22}$

15. $9\overline{)29}$

16. $6\overline{)21}$

17. $8\overline{)61}$

18. $9\overline{)64}$

19. $10\overline{)51}$

20. $11\overline{)40}$

21. $4\overline{)33}$

22. $6\overline{)57}$

23. $12\overline{)97}$

24. $8\overline{)60}$

Practice 15

Directions: Do these problems. Use your multiplication/division chart if you are unsure of your division facts. The first two are done for you. The next two are started for you.

1.
$$\begin{array}{r} 12\ \text{R1} \\ 3\overline{)37} \\ -3 \\ \hline 7 \\ -6 \\ \hline \end{array}$$

2.
$$\begin{array}{r} 13\ \text{R1} \\ 5\overline{)66} \\ -5 \\ \hline 16 \\ -15 \\ \hline \end{array}$$

3.
$$\begin{array}{r} 1\ \ \ \\ 3\overline{)48} \end{array}$$

4.
$$\begin{array}{r} 1\ \ \ \\ 4\overline{)59} \end{array}$$

5. $4\overline{)87}$

6. $3\overline{)69}$

7. $5\overline{)59}$

8. $7\overline{)84}$

9. $7\overline{)87}$

10. $9\overline{)93}$

11. $6\overline{)97}$

12. $5\overline{)61}$

13. $8\overline{)97}$

14. $3\overline{)72}$

15. $4\overline{)73}$

16. $6\overline{)79}$

17. $9\overline{)97}$

18. $3\overline{)87}$

19. $4\overline{)89}$

20. $6\overline{)81}$

Practice 16

Directions: Do these problems. Use your multiplication/division chart if you are unsure of your division facts. The first two are done for you.

1.
$$
\begin{array}{r}
41 \\
4\overline{)164} \\
-16 \\
\hline
4 \\
-4 \\
\hline
\end{array}
$$

2.
$$
\begin{array}{r}
51 \\
5\overline{)255} \\
-25 \\
\hline
5 \\
-5 \\
\hline
\end{array}
$$

3. $3\overline{)213}$

4. $3\overline{)363}$

5. $9\overline{)819}$

6. $6\overline{)366}$

7. $5\overline{)455}$

8. $3\overline{)333}$

9. $8\overline{)648}$

10. $2\overline{)466}$

11. $4\overline{)844}$

12. $6\overline{)666}$

13. $5\overline{)355}$

14. $7\overline{)357}$

15. $3\overline{)639}$

16. $2\overline{)248}$

17. $3\overline{)129}$

18. $4\overline{)244}$

19. $5\overline{)455}$

20. $3\overline{)936}$

Practice 17 ᗝ ᗣ ᗝ ᗣ ᗝ ᗣ ᗝ ᗣ ᗝ ᗣ ᗝ ᗣ ᗝ ᗣ ᗝ ᗣ

Directions: Do these problems. Use your multiplication/division chart if you are unsure of your division facts. The first two are done for you.

1.
$$
\begin{array}{r}
32 \\
6\overline{)192} \\
-18 \\
\hline
12 \\
-12 \\
\hline
\end{array}
$$

2.
$$
\begin{array}{r}
64 \\
4\overline{)256} \\
-24 \\
\hline
16 \\
-16 \\
\hline
\end{array}
$$

3. $4\overline{)224}$

4. $6\overline{)396}$

5. $4\overline{)804}$

6. $5\overline{)525}$

7. $4\overline{)828}$

8. $6\overline{)330}$

9. $9\overline{)945}$

10. $7\overline{)462}$

11. $3\overline{)309}$

12. $7\overline{)714}$

13. $4\overline{)392}$

14. $7\overline{)728}$

15. $9\overline{)666}$

16. $9\overline{)837}$

17. $8\overline{)944}$

18. $7\overline{)756}$

19. $6\overline{)354}$

20. $7\overline{)714}$

Practice 18 ೨ ೮ ೨ ೮ ೨ ೮ ೨ ೮ ೨ ೮ ೨ ೮ ೨ ೨ ೮

Directions: Do these problems. Use your multiplication/division chart if you are unsure of your division facts. The first one is done for you.

1.
$$
\begin{array}{r}
109 \\
4\overline{)436} \\
-4 \\
\hline
36 \\
-36 \\
\hline
\end{array}
$$

2. $5\overline{)615}$

3. $3\overline{)627}$

4. $3\overline{)318}$

5. $6\overline{)714}$

6. $8\overline{)368}$

7. $4\overline{)316}$

8. $3\overline{)432}$

9. $7\overline{)931}$

10. $5\overline{)655}$

11. $4\overline{)672}$

12. $8\overline{)912}$

13. $5\overline{)690}$

14. $6\overline{)822}$

15. $7\overline{)980}$

16. $4\overline{)948}$

17. $2\overline{)942}$

18. $6\overline{)858}$

Practice 19

Directions: Do these problems. Use your multiplication/division chart if you are unsure of your division facts. The first two are done for you.

1.
$$
\begin{array}{r}
71\ \text{R1} \\
3\overline{)214} \\
-21 \\
\hline
4 \\
-3 \\
\hline
1
\end{array}
$$

2.
$$
\begin{array}{r}
87\ \text{R1} \\
5\overline{)436} \\
-40 \\
\hline
36 \\
-35 \\
\hline
1
\end{array}
$$

3. $4\overline{)429}$

4. $4\overline{)517}$

5. $7\overline{)813}$

6. $6\overline{)297}$

7. $8\overline{)905}$

8. $3\overline{)776}$

9. $6\overline{)981}$

10. $5\overline{)759}$

11. $6\overline{)896}$

12. $7\overline{)818}$

13. $8\overline{)930}$

14. $7\overline{)944}$

15. $3\overline{)847}$

16. $5\overline{)631}$

17. $4\overline{)893}$

18. $3\overline{)574}$

Practice 20 ꙮ ꙮ ꙮ ꙮ ꙮ ꙮ ꙮ ꙮ ꙮ ꙮ ꙮ ꙮ ꙮ ꙮ ꙮ ꙮ

Directions: Do these problems. Use your multiplication/division chart if you are unsure of your division facts. The first two are done for you.

1.
$$
\begin{array}{r}
118\ R3 \\
7\overline{)829} \\
-7 \\
\hline
12 \\
-7 \\
\hline
59 \\
-56 \\
\hline
3
\end{array}
$$

2.
$$
\begin{array}{r}
184\ R1 \\
4\overline{)737} \\
-4 \\
\hline
33 \\
-32 \\
\hline
17 \\
-16 \\
\hline
1
\end{array}
$$

3. $8\overline{)197}$

4. $3\overline{)728}$

5. $6\overline{)718}$

6. $3\overline{)299}$

7. $4\overline{)977}$

8. $9\overline{)179}$

9. $5\overline{)948}$

10. $3\overline{)577}$

11. $2\overline{)111}$

12. $4\overline{)629}$

13. $3\overline{)703}$

14. $6\overline{)535}$

15. $7\overline{)978}$

16. $4\overline{)211}$

Practice 21 ⟳ ⟳ ⟳ ⟳ ⟳ ⟳ ⟳ ⟳ ⟳ ⟳ ⟳ ⟳ ⟳ ⟳

Directions: Do these problems. Use your multiplication/division chart if you are unsure of your division facts. The first two are done for you.

1.
$$\begin{array}{r} 612 \\ 6\overline{)3672} \\ -36 \\ \hline 7 \\ -6 \\ \hline 12 \\ -12 \\ \hline 0 \end{array}$$

2.
$$\begin{array}{r} 509 \\ 4\overline{)2036} \\ -20 \\ \hline 36 \\ -36 \\ \hline 0 \end{array}$$

3. $8\overline{)2408}$

4. $8\overline{)6416}$

5. $7\overline{)2107}$

6. $5\overline{)2535}$

7. $4\overline{)9024}$

8. $6\overline{)3006}$

9. $7\overline{)5047}$

10. $5\overline{)3005}$

11. $4\overline{)2712}$

12. $8\overline{)6488}$

13. $6\overline{)6006}$

14. $9\overline{)8118}$

15. $4\overline{)4000}$

Practice 22 ⟂ ⟂ ⟂ ⟂ ⟂ ⟂ ⟂ ⟂ ⟂ ⟂ ⟂ ⟂ ⟂ ⟂ ⟂

Directions: Do these problems. Use your multiplication/division chart if you are unsure of your division facts. The first two are done for you.

1.
```
      448 R1
  5)2241
   -20
    24
   -20
    41
   -40
     1
```

2.
```
      745 R1
  3)2236
   -21
    13
   -12
    16
   -15
     1
```

3. 7)3408

4. 5)2036

5. 3)2104

6. 6)3967

7. 5)4516

8. 7)3648

9. 9)6308

10. 7)1825

11. 3)4013

12. 6)1761

13. 7)2019

14. 3)1338

15. 7)3211

16. 4)8887

Practice 23 ✇ ✇ ✇ ✇ ✇ ✇ ✇ ✇ ✇ ✇ ✇ ✇ ✇ ✇ ✇

Reminders

- A dividend is divisible by a divisor if it can be divided evenly by that divisor with no remainder.
- Any number ending in 5 or 0 is divisible by 5.
- Any number ending in 0 is divisible by 10.

Directions: Do these problems. Use your multiplication/division chart if you are unsure of your division facts. The first two are done for you.

1.
```
      81
  5)405
  -40
    5
   -5
    0
```

2.
```
     104
  5)520
  -50
    20
   -20
    0
```

3. 10)100

4. 5)215

5. 10)720

6. 10)140

7. 5)680

8. 10)220

9. 10)290

10. 5)620

11. 10)440

12. 10)990

13. 10)4780

14. 5)1655

15. 10)4330

16. 10)3210

Practice 24

Directions: Do these problems. Use your multiplication/division chart if you are unsure of your division facts. The first two are done for you.

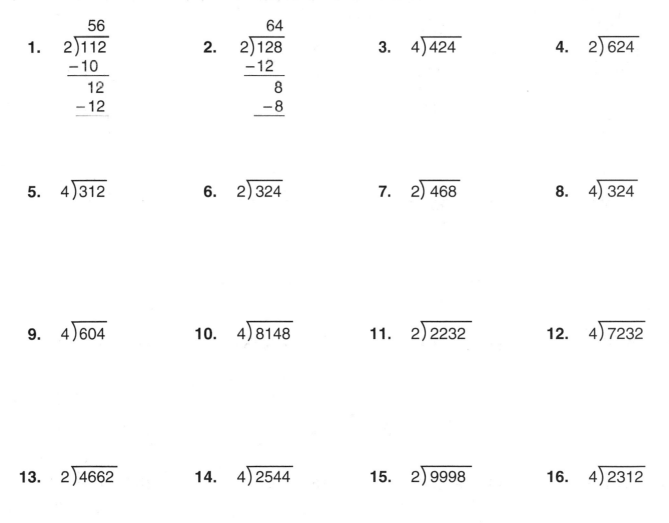

1. $\begin{array}{r} 56 \\ 2\overline{)112} \\ -10 \\ \hline 12 \\ -12 \end{array}$

2. $\begin{array}{r} 64 \\ 2\overline{)128} \\ -12 \\ \hline 8 \\ -8 \end{array}$

3. $4\overline{)424}$

4. $2\overline{)624}$

5. $4\overline{)312}$

6. $2\overline{)324}$

7. $2\overline{)468}$

8. $4\overline{)324}$

9. $4\overline{)604}$

10. $4\overline{)8148}$

11. $2\overline{)2232}$

12. $4\overline{)7232}$

13. $2\overline{)4662}$

14. $4\overline{)2544}$

15. $2\overline{)9998}$

16. $4\overline{)2312}$

Practice 25 ༄ ༄ ༄ ༄ ༄ ༄ ༄ ༄ ༄ ༄ ༄ ༄ ༄ ༄ ༄

Reminders

- A dividend is divisible by a divisor if it can be divided evenly by that divisor with no remainder.
- A dividend is divisible by 9 if the sum of the digits in the dividend equal 9 or a multiple of 9, such as 18, 27, 36, etc.

Example: $9\overline{)4572}$ ➤ The dividend 4572 is divisible by 9 because the sum of $4 + 5 + 7 + 2 = 18$.

Directions: Do these problems. Use your multiplication/division chart if you are unsure of your division facts. The first one is done for you.

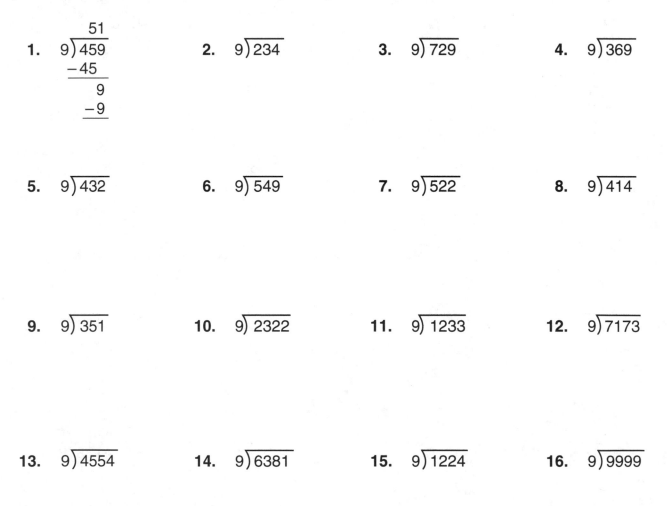

$$
\begin{array}{r}
51 \\
1. \quad 9\overline{)459} \\
-45 \\
\hline
9 \\
-9 \\
\hline
\end{array}
$$

2. $9\overline{)234}$

3. $9\overline{)729}$

4. $9\overline{)369}$

5. $9\overline{)432}$

6. $9\overline{)549}$

7. $9\overline{)522}$

8. $9\overline{)414}$

9. $9\overline{)351}$

10. $9\overline{)2322}$

11. $9\overline{)1233}$

12. $9\overline{)7173}$

13. $9\overline{)4554}$

14. $9\overline{)6381}$

15. $9\overline{)1224}$

16. $9\overline{)9999}$

Practice 26 ◑ ◕ ◑ ◑ ◑ ◕ ◑ ◕ ◑ ◑ ◕ ◑ ◑ ◑ ◕

Directions: Do these problems. Use your multiplication/division chart if you are unsure of your division facts. The first one is done for you.

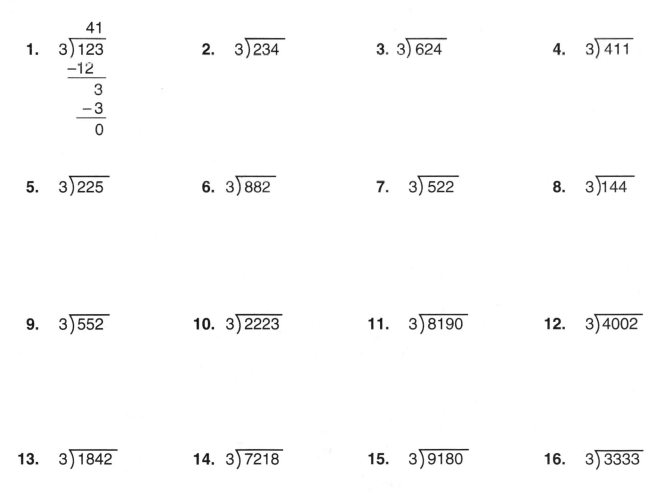

1.
$$\begin{array}{r} 41 \\ 3\overline{)123} \\ -12 \\ \hline 3 \\ -3 \\ \hline 0 \end{array}$$

2. $3\overline{)234}$

3. $3\overline{)624}$

4. $3\overline{)411}$

5. $3\overline{)225}$

6. $3\overline{)882}$

7. $3\overline{)522}$

8. $3\overline{)144}$

9. $3\overline{)552}$

10. $3\overline{)2223}$

11. $3\overline{)8190}$

12. $3\overline{)4002}$

13. $3\overline{)1842}$

14. $3\overline{)7218}$

15. $3\overline{)9180}$

16. $3\overline{)3333}$

Practice 27 ⟫ ◉ ⟫ ◉ ◉ ⟫ ◉ ◉ ⟫ ◉ ◉ ⟫ ◉ ⟫ ⟫ ◉

Reminders

- A dividend is divisible by a divisor if it can be divided evenly by that divisor with no remainder.
- Any number ending in 0 is divisible by 10.
- Any number ending in 00 is divisible by 100.

Directions: Do these problems. Use your multiplication/division chart if you are unsure of your division facts. The first one is done for you.

1.
$$
\begin{array}{r}
42 \\
10\overline{)420} \\
-40 \\
\hline
20 \\
-20 \\
\hline
0
\end{array}
$$

2. $10\overline{)660}$

3. $10\overline{)210}$

4. $10\overline{)280}$

5. $10\overline{)950}$

6. $10\overline{)940}$

7. $100\overline{)900}$

8. $100\overline{)700}$

9. $100\overline{)800}$

10. $10\overline{)7660}$

11. $10\overline{)4340}$

12. $10\overline{)7750}$

13. $100\overline{)8700}$

14. $100\overline{)6600}$

15. $100\overline{)4300}$

16. $100\overline{)5600}$

Practice 28 ⟡ ⟡ ⟡ ⟡ ⟡ ⟡ ⟡ ⟡ ⟡ ⟡ ⟡ ⟡ ⟡ ⟡

Reminders

- A dividend is divisible by a divisor if it can be divided evenly by that divisor with no remainder.
- Any number ending in 00, 20, 40, 60, or 80 is divisible by 20.
- Any number ending in 00, 25, 50, or 75 is divisible by 25.

Directions: Do these problems. Use your multiplication/division chart if you are unsure of your division facts. The first one is done for you.

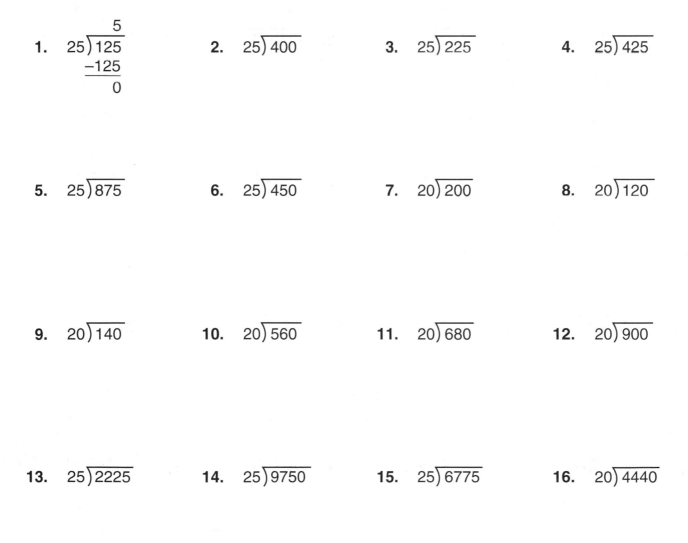

1. $25\overline{)125}$
   $\;\;5$
   -125
   $\;\;\;0$

2. $25\overline{)400}$

3. $25\overline{)225}$

4. $25\overline{)425}$

5. $25\overline{)875}$

6. $25\overline{)450}$

7. $20\overline{)200}$

8. $20\overline{)120}$

9. $20\overline{)140}$

10. $20\overline{)560}$

11. $20\overline{)680}$

12. $20\overline{)900}$

13. $25\overline{)2225}$

14. $25\overline{)9750}$

15. $25\overline{)6775}$

16. $20\overline{)4440}$

Practice 29 ⬧ ⬧ ⬧ ⬧ ⬧ ⬧ ⬧ ⬧ ⬧ ⬧ ⬧ ⬧

Directions: Do these problems. Don't forget to use dollar signs and decimal points. Use your multiplication/division chart if you are unsure of your division facts. The first one is done for you.

1.
$$
\begin{array}{r}
\$.57 \\
5\overline{)\$2.85} \\
-2\,5 \\
\hline
35 \\
-35 \\
\hline
0
\end{array}
$$

2. $4\overline{)\$1.24}$

3. $3\overline{)\$3.03}$

4. $2\overline{)\$2.10}$

5. $6\overline{)\$1.20}$

6. $9\overline{)\$9.18}$

7. $4\overline{)\$8.80}$

8. $7\overline{)\$1.40}$

9. $3\overline{)\$2.13}$

10. $5\overline{)\$9.25}$

11. $9\overline{)\$4.50}$

12. $2\overline{)\$7.18}$

13. $7\overline{)\$7.35}$

14. $3\overline{)\$1.83}$

15. $4\overline{)\$9.96}$

16. $9\overline{)\$9.36}$

Practice 30 ꩜ ꩜ ꩜ ꩜ ꩜ ꩜ ꩜ ꩜ ꩜ ꩜ ꩜ ꩜ ꩜ ꩜

Directions: Do these problems. Don't forget to use dollar signs and decimal points. Use your multiplication/division chart if you are unsure of your division facts. The first one is done for you.

1.
$$\begin{array}{r} \$\ 5.05 \\ 5\overline{)\$25.25} \\ -25 \\ \hline 25 \\ -25 \\ \hline \end{array}$$

2. $2\overline{)\$12.44}$

3. $4\overline{)\$88.44}$

4. $6\overline{)\$36.18}$

5. $7\overline{)\$12.60}$

6. $9\overline{)\$28.44}$

7. $2\overline{)\$28.28}$

8. $5\overline{)\$29.75}$

9. $3\overline{)\$13.29}$

10. $8\overline{)\$41.68}$

11. $7\overline{)\$41.30}$

12. $6\overline{)\$58.80}$

13. $5\overline{)\$98.45}$

14. $3\overline{)\$71.10}$

15. $2\overline{)\$71.98}$

16. $3\overline{)\$17.13}$

Practice 31 ﹥ ❂ ﹥ ❂ ﹥ ❂ ﹥ ❂ ﹥ ❂ ﹥ ❂ ﹥ ❂ ﹥ ﹥ ❂

Directions: Do these problems. Use your multiplication/division chart if you are unsure of your division facts. The first two are done for you.

1.
$$
\begin{array}{r}
6 \\
20\overline{)120} \\
-120 \\
\hline
0
\end{array}
$$

2.
$$
\begin{array}{r}
11 \\
20\overline{)220} \\
-20 \\
\hline
20 \\
-20 \\
\hline
0
\end{array}
$$

3. $20\overline{)340}$

4. $30\overline{)270}$

5. $30\overline{)180}$

6. $30\overline{)360}$

7. $20\overline{)480}$

8. $30\overline{)330}$

9. $20\overline{)960}$

10. $40\overline{)440}$

11. $40\overline{)880}$

12. $50\overline{)250}$

13. $20\overline{)760}$

14. $40\overline{)560}$

15. $50\overline{)750}$

16. $30\overline{)780}$

Practice 32 ⟳ ⟳ ⟳ ⟳ ⟳ ⟳ ⟳ ⟳ ⟳ ⟳ ⟳ ⟳ ⟳ ⟳

Directions: Do these problems. Use your multiplication/division chart if you are unsure of your division facts. The first two are done for you.

1.
$$\begin{array}{r} 8 \\ 20\overline{)160} \\ -160 \\ \hline 0 \end{array}$$

2.
$$\begin{array}{r} 7 \\ 30\overline{)210} \\ -210 \\ \hline 0 \end{array}$$

3. $40\overline{)240}$

4. $20\overline{)280}$

5. $40\overline{)160}$

6. $50\overline{)350}$

7. $40\overline{)960}$

8. $20\overline{)380}$

9. $30\overline{)150}$

10. $40\overline{)640}$

11. $40\overline{)800}$

12. $50\overline{)750}$

13. $30\overline{)270}$

14. $30\overline{)660}$

15. $30\overline{)990}$

16. $40\overline{)520}$

Practice 33 ๑ ๏ ๏ ๑ ๏ ๏ ๑ ๏ ๑ ๑ ๏ ๏ ๑ ๏ ๑ ๑ ๏

Directions: Do these problems. Use your multiplication/division chart if you are unsure of your division facts. The first two are done for you.

1.
$$\begin{array}{r} 6\ R10 \\ 20\overline{)130} \\ -120 \\ \hline 10 \end{array}$$

2.
$$\begin{array}{r} 4\ R20 \\ 30\overline{)140} \\ -120 \\ \hline 20 \end{array}$$

3. $40\overline{)150}$

4. $40\overline{)290}$

5. $50\overline{)180}$

6. $20\overline{)250}$

7. $20\overline{)170}$

8. $20\overline{)230}$

9. $30\overline{)190}$

10. $50\overline{)140}$

11. $50\overline{)220}$

12. $40\overline{)350}$

13. $20\overline{)210}$

14. $40\overline{)330}$

15. $20\overline{)190}$

16. $40\overline{)130}$

Practice 34 ⟳ ⟲ ⟳ ⟲ ⟳ ⟲ ⟳ ⟲ ⟳ ⟲ ⟳ ⟲ ⟳ ⟳ ⟲

Directions: Do these problems. Use your multiplication/division chart if you are unsure of your division facts. The first one is done for you.

1.
$$
\begin{array}{r}
14\ R10 \\
20\,\overline{)290} \\
-20 \\
\hline
90 \\
-80 \\
\hline
10
\end{array}
$$

2. $30\,\overline{)370}$

3. $40\,\overline{)570}$

4. $30\,\overline{)590}$

5. $40\,\overline{)310}$

6. $20\,\overline{)970}$

7. $30\,\overline{)520}$

8. $40\,\overline{)830}$

9. $50\,\overline{)690}$

10. $20\,\overline{)630}$

11. $50\,\overline{)440}$

12. $40\,\overline{)910}$

13. $40\,\overline{)410}$

14. $50\,\overline{)730}$

15. $20\,\overline{)990}$

16. $50\,\overline{)535}$

Practice 35 ⤧ ❂ ⤧ ❂ ⤧ ❂ ⤧ ❂ ⤧ ❂ ❂ ⤧ ❂ ⤧ ⤧ ❂

Directions: Use your division skills to solve these word problems. Use your multiplication/division chart, if needed.

1. What is 155 divided by 5? _____

2. What is the quotient when 144 is divided by 9? _____

3. Janna has a bag containing 250 candy rolls. She is going to split them evenly among 10 friends. How many candy rolls will each friend receive? _____

4. Divide 650 by 25. _____

5. Bernie found 489 pennies in an old piggy bank. He is going to arrange the pennies evenly in 3 separate piles. How many pennies will he put in each pile? _____

6. The divisor is 7. The dividend is 413. What is the quotient? _____

7. What is 376 divided by 4? _____

8. Sarah's teacher told her to divide 420 strips of paper evenly among the 30 students in her class. How many strips of paper did each student receive? _____

9. Find the quotient for 999 ÷ 3. _____

10. Divide 288 by 4. _____

11. James had to split a bag of 85 lollipops among 10 children at a party.
    How many lollipops did each child receive? _____
    How many lollipops were left over? _____

12. What is 342 divided by 9? _____

Practice 36 ᔐ ᘓ ᔐ ᘓ ᔐ ᘓ ᔐ ᘓ ᔐ ᘓ ᔐ ᘓ ᔐ ᔐ ᘓ

Directions: Use your division skills to solve these word problems. Use your multiplication/division chart, if needed.

1. Jason had 156 baseball cards which he arranged in 4 equal piles. How many cards were in each pile?_____

2. Daniel dealt a deck of 52 cards to 4 players. How many cards did each player receive?

    _____

3. What is the quotient when 288 is divided by 9? _____

4. What is 698 divided by 4? _____

    What is the remainder? _____

5. Bruce found 960 seeds in a pumpkin. He divided them evenly among the 30 students in his class. How many seeds did each student receive? _____

6. Divide 975 by 25. _____

7. The divisor is 9. The dividend is 324. What is the quotient? _____

8. What is 987 divided by 7?_____

9. Michael's teacher asked him to split 880 kernels of corn among 20 students for a science experiment. How many kernels of corn did each student receive? _____

10. Susan had to cut a length of masking tape, which was 70 inches long into pieces which were each 5 inches long. How many pieces did she cut? _____

11. Find the quotient: 578 divided by 2. _____

12. Divide 770 by 8. _____

Test Practice 1

1.

$4\overline{)52}$

- (A) 12
- (B) 13
- (C) 15
- (D) 23

2.

$8\overline{)88}$

- (A) 18
- (B) 12
- (C) 11
- (D) 9

3.

$3\overline{)39}$

- (A) 19
- (B) 12
- (C) 13
- (D) 11

4.

$5\overline{)70}$

- (A) 12
- (B) 140
- (C) 13
- (D) 14

5.

$4\overline{)44}$

- (A) 12
- (B) 11
- (C) 13
- (D) 111

6.

$6\overline{)84}$

- (A) 16
- (B) 14
- (C) 13
- (D) 15

7.

$7\overline{)91}$

- (A) 14
- (B) 11
- (C) 13
- (D) 12

8.

$8\overline{)96}$

- (A) 12
- (B) 14
- (C) 15
- (D) 13

9.

$6\overline{)36}$

- (A) 6
- (B) 12
- (C) 7
- (D) 13

10.

$9\overline{)99}$

- (A) 11
- (B) 9
- (C) 13
- (D) 14

11.

$6 \times \underline{} = 66$

- (A) 10
- (B) 12
- (C) 11
- (D) 13

12.

$9 \times \underline{} = 72$

- (A) 9
- (B) 8
- (C) 7
- (D) 12

13.

$\underline{} \times 9 = 90$

- (A) 10
- (B) 9
- (C) 11
- (D) 8

14.

$8 \times \underline{} = 56$

- (A) 12
- (B) 7
- (C) 8
- (D) 9

Test Practice 2 ⟡⟡⟡⟡⟡⟡⟡⟡⟡⟡⟡

1.
$9\overline{)88}$
Ⓐ 9
Ⓑ 9 R8
Ⓒ 9 R6
Ⓓ 9 R7

8.
$6\overline{)77}$
Ⓐ 12 R3
Ⓑ 11 R4
Ⓒ 12 R5
Ⓓ 12 R6

2.
$6\overline{)91}$
Ⓐ 15
Ⓑ 15 R1
Ⓒ 15 R3
Ⓓ 16 R1

9.
$4\overline{)87}$
Ⓐ 21 R3
Ⓑ 22 R3
Ⓒ 20 R7
Ⓓ 21 R1

3.
$4\overline{)15}$
Ⓐ 3 R3
Ⓑ 3 R4
Ⓒ 4 R1
Ⓓ 2 R5

10.
$6\overline{)41}$
Ⓐ 6 R4
Ⓑ 6 R5
Ⓒ 7 R1
Ⓓ 6 R3

4.
$7\overline{)67}$
Ⓐ 9 R3
Ⓑ 9 R5
Ⓒ 9 R6
Ⓓ 9 R4

11.
$6\overline{)88}$
Ⓐ 6 R7
Ⓑ 14 R4
Ⓒ 6 R9
Ⓓ 14 R2

5.
$5\overline{)49}$
Ⓐ 9 R3
Ⓑ 8 R4
Ⓒ 9 R5
Ⓓ 9 R4

12.
$5\overline{)87}$
Ⓐ 17 R3
Ⓑ 16 R3
Ⓒ 17 R2
Ⓓ 16 R7

6.
$6\overline{)50}$
Ⓐ 8 R2
Ⓑ 8 R4
Ⓒ 9 R5
Ⓓ 9 R6

13.
$8\overline{)71}$
Ⓐ 8 R6
Ⓑ 8 R7
Ⓒ 9 R1
Ⓓ 8 R8

7.
$5\overline{)93}$
Ⓐ 16 R3
Ⓑ 18 R4
Ⓒ 17 R3
Ⓓ 18 R3

14.
$3\overline{)56}$
Ⓐ 19 R2
Ⓑ 18 R2
Ⓒ 18 R1
Ⓓ 15 R2

Test Practice 3 ᓂ ᘒ ᓂ ᘒ ᓂ ᘒ ᘒ ᓂ ᘒ ᓂ ᓂ ᘒ

1.
$5\overline{)125}$
- (A) 23
- (B) 25
- (C) 30
- (D) 15

2.
$7\overline{)217}$
- (A) 33
- (B) 37
- (C) 31
- (D) 21

3.
$6\overline{)246}$
- (A) 31
- (B) 42
- (C) 41
- (D) 43

4.
$8\overline{)656}$
- (A) 88
- (B) 62
- (C) 83
- (D) 82

5.
$7\overline{)154}$
- (A) 21
- (B) 22
- (C) 23
- (D) 24

6.
$4\overline{)268}$
- (A) 68
- (B) 69
- (C) 66
- (D) 67

7.
$9\overline{)198}$
- (A) 23
- (B) 21
- (C) 32
- (D) 22

8.
$3\overline{)432}$
- (A) 144
- (B) 164
- (C) 244
- (D) 154

9.
$5\overline{)755}$
- (A) 255
- (B) 205
- (C) 151
- (D) 155

10.
$7\overline{)889}$
- (A) 172
- (B) 127
- (C) 129
- (D) 117

11.
$4\overline{)904}$
- (A) 225
- (B) 215
- (C) 216
- (D) 226

12.
$6\overline{)606}$
- (A) 111
- (B) 101
- (C) 110
- (D) 121

13.
$4\overline{)812}$
- (A) 230
- (B) 233
- (C) 213
- (D) 203

14.
$3\overline{)930}$
- (A) 310
- (B) 301
- (C) 330
- (D) 311

Test Practice 4 ᔕ ᕮ ᔕ ᕮ ᔕ ᕮ ᔕ ᕮ ᔕ ᔕ ᕮ

1.

$5\overline{)180}$

(A) 30
(B) 838
(C) 36
(D) 39

2.

$10\overline{)410}$

(A) 41
(B) 401
(C) 14
(D) 410

3.

$5\overline{)915}$

(A) 180
(B) 185
(C) 18
(D) 183

4.

$25\overline{)625}$

(A) 21
(B) 225
(C) 26
(D) 25

5.

$2\overline{)134}$

(A) 67
(B) 64
(C) 61
(D) 66

6.

$4\overline{)924}$

(A) 23
(B) 213
(C) 206
(D) 231

7.

$9\overline{)243}$

(A) 25
(B) 26
(C) 37
(D) 27

8.

$9\overline{)558}$

(A) 62
(B) 63
(C) 72
(D) 61

9.

$3\overline{)126}$

(A) 44
(B) 43
(C) 52
(D) 42

10.

$3\overline{)423}$

(A) 142
(B) 141
(C) 151
(D) 143

11.

$5\overline{)805}$

(A) 162
(B) 163
(C) 160
(D) 161

12.

$20\overline{)940}$

(A) 407
(B) 46
(C) 47
(D) 417

13.

$2\overline{)866}$

(A) 443
(B) 423
(C) 432
(D) 433

14.

$10\overline{)630}$

(A) 630
(B) 63
(C) 603
(D) 613

Test Practice 5

1.

$4\overline{)139}$

- (A) 34 R2
- (B) 34 R3
- (C) 35 R3
- (D) 34 R1

2.

$3\overline{)167}$

- (A) 55 R1
- (B) 55 R2
- (C) 45 R2
- (D) 65 R2

3.

$5\overline{)179}$

- (A) 35 R4
- (B) 34 R4
- (C) 60 R4
- (D) 35 R3

4.

$7\overline{)257}$

- (A) 36
- (B) 36 R5
- (C) 36 R2
- (D) 37 R2

5.

$8\overline{)281}$

- (A) 35 R3
- (B) 35 R1
- (C) 36 R1
- (D) 37 R1

6.

$8\overline{)407}$

- (A) 50 R6
- (B) 50 R7
- (C) 51 R7
- (D) 55 R7

7.

$3\overline{)632}$

- (A) 211 R2
- (B) 210 R1
- (C) 212 R2
- (D) 210 R2

8.

$7\overline{)564}$

- (A) 80 R5
- (B) 81 R5
- (C) 80 R4
- (D) 81 R4

9.

$9\overline{)906}$

- (A) 100 R6
- (B) 111 R6
- (C) 100 R3
- (D) 100 R7

10.

$2\overline{)763}$

- (A) 382 R1
- (B) 371 R1
- (C) 341 R1
- (D) 381 R1

11.

$6\overline{)671}$

- (A) 111 R4
- (B) 101 R5
- (C) 111 R3
- (D) 111 R5

12.

$8\overline{)991}$

- (A) 122 R6
- (B) 123 R6
- (C) 123 R7
- (D) 133 R7

13.

$3\overline{)311}$

- (A) 104 R2
- (B) 102 R2
- (C) 103 R1
- (D) 103 R2

14.

$8\overline{)473}$

- (A) 69 R1
- (B) 59 R5
- (C) 63 R6
- (D) 59 R1

#3323 Practice Makes Perfect: Division

Test Practice 6

1. $20\overline{)160}$
- (A) 9
- (B) 8
- (C) 7
- (D) 6

2. $30\overline{)390}$
- (A) 33
- (B) 16
- (C) 13
- (D) 23

3. $40\overline{)280}$
- (A) 6
- (B) 80
- (C) 70
- (D) 7

4. $50\overline{)650}$
- (A) 15
- (B) 19
- (C) 13
- (D) 130

5. $30\overline{)990}$
- (A) 31
- (B) 39
- (C) 38
- (D) 33

6. $40\overline{)920}$
- (A) 22
- (B) 23
- (C) 20
- (D) 24

7. $30\overline{)140}$
- (A) 40 R20
- (B) 4 R10
- (C) 4 R20
- (D) 3 R20

8. $50\overline{)710}$
- (A) 14 R10
- (B) 140 R10
- (C) 14 R10
- (D) 14 R20

9. $40\overline{)270}$
- (A) 5 R20
- (B) 6 R20
- (C) 5 R30
- (D) 6 R30

10. $50\overline{)430}$
- (A) 8 R20
- (B) 9 R20
- (C) 8 R30
- (D) 80 R30

11. $30\overline{)490}$
- (A) 6 R10
- (B) 18 R10
- (C) 16 R20
- (D) 16 R10

12. $20\overline{)365}$
- (A) 18 R15
- (B) 18 R5
- (C) 9 R15
- (D) 16 R15

13. $40\overline{)804}$
- (A) 2 R4
- (B) 21 R4
- (C) 41 R4
- (D) 20 R4

14. $20\overline{)629}$
- (A) 30 R19
- (B) 31 R19
- (C) 31 R9
- (D) 30 R9

Answer Sheet

Test Practice 1

1. Ⓐ Ⓑ Ⓒ Ⓓ
2. Ⓐ Ⓑ Ⓒ Ⓓ
3. Ⓐ Ⓑ Ⓒ Ⓓ
4. Ⓐ Ⓑ Ⓒ Ⓓ
5. Ⓐ Ⓑ Ⓒ Ⓓ
6. Ⓐ Ⓑ Ⓒ Ⓓ
7. Ⓐ Ⓑ Ⓒ Ⓓ
8. Ⓐ Ⓑ Ⓒ Ⓓ
9. Ⓐ Ⓑ Ⓒ Ⓓ
10. Ⓐ Ⓑ Ⓒ Ⓓ
11. Ⓐ Ⓑ Ⓒ Ⓓ
12. Ⓐ Ⓑ Ⓒ Ⓓ
13. Ⓐ Ⓑ Ⓒ Ⓓ
14. Ⓐ Ⓑ Ⓒ Ⓓ

Test Practice 2

1. Ⓐ Ⓑ Ⓒ Ⓓ
2. Ⓐ Ⓑ Ⓒ Ⓓ
3. Ⓐ Ⓑ Ⓒ Ⓓ
4. Ⓐ Ⓑ Ⓒ Ⓓ
5. Ⓐ Ⓑ Ⓒ Ⓓ
6. Ⓐ Ⓑ Ⓒ Ⓓ
7. Ⓐ Ⓑ Ⓒ Ⓓ
8. Ⓐ Ⓑ Ⓒ Ⓓ
9. Ⓐ Ⓑ Ⓒ Ⓓ
10. Ⓐ Ⓑ Ⓒ Ⓓ
11. Ⓐ Ⓑ Ⓒ Ⓓ
12. Ⓐ Ⓑ Ⓒ Ⓓ
13. Ⓐ Ⓑ Ⓒ Ⓓ
14. Ⓐ Ⓑ Ⓒ Ⓓ

Test Practice 3

1. Ⓐ Ⓑ Ⓒ Ⓓ
2. Ⓐ Ⓑ Ⓒ Ⓓ
3. Ⓐ Ⓑ Ⓒ Ⓓ
4. Ⓐ Ⓑ Ⓒ Ⓓ
5. Ⓐ Ⓑ Ⓒ Ⓓ
6. Ⓐ Ⓑ Ⓒ Ⓓ
7. Ⓐ Ⓑ Ⓒ Ⓓ
8. Ⓐ Ⓑ Ⓒ Ⓓ
9. Ⓐ Ⓑ Ⓒ Ⓓ
10. Ⓐ Ⓑ Ⓒ Ⓓ
11. Ⓐ Ⓑ Ⓒ Ⓓ
12. Ⓐ Ⓑ Ⓒ Ⓓ
13. Ⓐ Ⓑ Ⓒ Ⓓ
14. Ⓐ Ⓑ Ⓒ Ⓓ

Test Practice 4

1. Ⓐ Ⓑ Ⓒ Ⓓ
2. Ⓐ Ⓑ Ⓒ Ⓓ
3. Ⓐ Ⓑ Ⓒ Ⓓ
4. Ⓐ Ⓑ Ⓒ Ⓓ
5. Ⓐ Ⓑ Ⓒ Ⓓ
6. Ⓐ Ⓑ Ⓒ Ⓓ
7. Ⓐ Ⓑ Ⓒ Ⓓ
8. Ⓐ Ⓑ Ⓒ Ⓓ
9. Ⓐ Ⓑ Ⓒ Ⓓ
10. Ⓐ Ⓑ Ⓒ Ⓓ
11. Ⓐ Ⓑ Ⓒ Ⓓ
12. Ⓐ Ⓑ Ⓒ Ⓓ
13. Ⓐ Ⓑ Ⓒ Ⓓ
14. Ⓐ Ⓑ Ⓒ Ⓓ

Test Practice 5

1. Ⓐ Ⓑ Ⓒ Ⓓ
2. Ⓐ Ⓑ Ⓒ Ⓓ
3. Ⓐ Ⓑ Ⓒ Ⓓ
4. Ⓐ Ⓑ Ⓒ Ⓓ
5. Ⓐ Ⓑ Ⓒ Ⓓ
6. Ⓐ Ⓑ Ⓒ Ⓓ
7. Ⓐ Ⓑ Ⓒ Ⓓ
8. Ⓐ Ⓑ Ⓒ Ⓓ
9. Ⓐ Ⓑ Ⓒ Ⓓ
10. Ⓐ Ⓑ Ⓒ Ⓓ
11. Ⓐ Ⓑ Ⓒ Ⓓ
12. Ⓐ Ⓑ Ⓒ Ⓓ
13. Ⓐ Ⓑ Ⓒ Ⓓ
14. Ⓐ Ⓑ Ⓒ Ⓓ

Test Practice 6

1. Ⓐ Ⓑ Ⓒ Ⓓ
2. Ⓐ Ⓑ Ⓒ Ⓓ
3. Ⓐ Ⓑ Ⓒ Ⓓ
4. Ⓐ Ⓑ Ⓒ Ⓓ
5. Ⓐ Ⓑ Ⓒ Ⓓ
6. Ⓐ Ⓑ Ⓒ Ⓓ
7. Ⓐ Ⓑ Ⓒ Ⓓ
8. Ⓐ Ⓑ Ⓒ Ⓓ
9. Ⓐ Ⓑ Ⓒ Ⓓ
10. Ⓐ Ⓑ Ⓒ Ⓓ
11. Ⓐ Ⓑ Ⓒ Ⓓ
12. Ⓐ Ⓑ Ⓒ Ⓓ
13. Ⓐ Ⓑ Ⓒ Ⓓ
14. Ⓐ Ⓑ Ⓒ Ⓓ

Answer Key ⟆ ◉ ⟆ ◉ ⟆ ◉ ⟆ ◉ ⟆ ◉ ⟆ ◉ ⟆ ⟆ ◉

Page 4
1. 21, 18, 15, 12, 9, 6, 3
2. 12, 10, 8, 6, 4, 2
3. 32, 28, 24, 20, 16, 12, 8, 4
4. 35, 30, 25, 20, 15, 10, 5
5. 21, 14, 7
6. 16, 8
7. 48, 36, 24, 12
8. the 10's

Page 5
1. 28, 24, 20, 16, 12, 8, 4
2. 49, 42, 35, 28, 21, 14, 7
3. 35, 30, 25, 20, 15, 10, 5
4. 70, 60, 50, 40, 30, 20, 10
5. 7, 6, 5, 4, 3, 2, 1
6. the 10's
7. six
8. four
9. six
10. four
11. five
12. six

Page 6
1. 5
2. 5
3. 1
4. 4
5. 1
6. 4
7. 3
8. 3
9. 6
10. 4
11. 3
12. 5
13. 7
14. 7
15. 4
16. 9
17. 7
18. 8
19. 4
20. 10
21. 9
22. 7
23. 8
24. 10
25. 8
26. 9
27. 10
28. 6
29. 2
30. 9
31. 10
32. 2
33. 5
34. 4
35. 4
36. 9
37. 5
38. 6
39. 11
40. 8

Page 7
1. 2
2. 2
3. 2
4. 4
5. 3
6. 3
7. 4
8. 9
9. 5
10. 7
11. 12
12. 8
13. 8
14. 6
15. 9
16. 12
17. 9
18. 7
19. 1
20. 5
21. 10
22. 1
23. 4
24. 9
25. 2
26. 3
27. 4
28. 5
29. 7
30. 11
31. 12
32. 10
33. 3
34. 12
35. 8
36. 6
37. 9
38. 10
39. 11
40. 12

Page 8
1. 2
2. 3
3. 2
4. 2
5. 3
6. 5
7. 3
8. 4
9. 5
10. 4
11. 9
12. 6
13. 4
14. 6
15. 8
16. 8
17. 7
18. 6
19. 1
20. 7
21. 7
22. 8
23. 7
24. 12
25. 9
26. 7
27. 8
28. 5
29. 10
30. 11
31. 12
32. 5
33. 1
34. 4
35. 7
36. 4
37. 6
38. 8
39. 9
40. 9

Page 9
1. 1
2. 2
3. 2
4. 1
5. 3
6. 5
7. 5
8. 9
9. 4
10. 7
11. 9
12. 9
13. 8
14. 8
15. 3
16. 7
17. 4
18. 12
19. 10
20. 11
21. 11
22. 6
23. 8
24. 4
25. 4
26. 6
27. 8
28. 8
29. 5
30. 3
31. 5
32. 11
33. 2
34. 8
35. 7
36. 9
37. 10
38. 9
39. 12
40. 4

Page 10
1. 2
2. 4
3. 3
4. 5
5. 4
6. 6
7. 6
8. 7
9. 4
10. 2
11. 7
12. 8
13. 6
14. 9
15. 10
16. 11
17. 9
18. 7
19. 9
20. 8
21. 8
22. 10
23. 12
24. 11
25. 2
26. 6
27. 12
28. 6
29. 12
30. 12
31. 5
32. 6
33. 3
34. 4
35. 11
36. 7
37. 5
38. 6
39. 10
40. 8

Page 11
1. 2
2. 5
3. 3
4. 3
5. 4
6. 6
7. 3
8. 9
9. 5
10. 6
11. 6
12. 10
13. 7
14. 4
15. 2

Page 12
16. 11
17. 5
18. 3
19. 8
20. 10
21. 2
22. 4
23. 8
24. 9
25. 6
26. 3
27. 7
28. 4
29. 8
30. 11
31. 4
32. 7
33. 1
34. 4
35. 5
36. 12
37. 5
38. 7
39. 8
40. 12

Page 12
1. 4
2. 5
3. 2
4. 6
5. 3
6. 7
7. 8
8. 3
9. 6
10. 11
11. 7
12. 11
13. 12
14. 6
15. 6
16. 5
17. 6
18. 9
19. 12
20. 5
21. 4
22. 8
23. 5
24. 3
25. 8
26. 6
27. 3
28. 2
29. 10
30. 9
31. 10
32. 12
33. 3
34. 9
35. 7
36. 8
37. 9
38. 8
39. 5
40. 9

Page 13
1. 10
2. 8
3. 9
4. 9
5. 5
6. 9
7. 7
8. 7
9. 11
10. 4
11. 7

Page 14
12. 9
13. 6
14. 11
15. 9
16. 6
17. 9
18. 7
19. 12
20. 9
21. 6
22. 9
23. 3
24. 9
25. 9
26. 6
27. 4
28. 3
29. 8
30. 6
31. 9
32. 12
33. 9
34. 2
35. 7
36. 4
37. 6
38. 8
39. 4
40. 12

Page 14
1. 6
2. 8
3. 4
4. 7
5. 9
6. 9
7. 9
8. 4
9. 6
10. 6
11. 7
12. 7
13. 8
14. 7
15. 8
16. 9
17. 6
18. 9
19. 9
20. 6
21. 4
22. 8
23. 9
24. 3
25. 7
26. 5
27. 7
28. 12
29. 6
30. 7
31. 11
32. 12
33. 11
34. 9
35. 6
36. 4

Page 15
1. 5 R1
2. 4 R2
3. 4 R1
4. 6 R1
5. 9 R1
6. 5 R1
7. 7 R1
8. 4 R1

9. 8 R3
10. 7 R1
11. 7 R1
12. 4 R3
13. 8 R1
14. 7 R1
15. 9 R3
16. 7 R1
17. 9 R1
18. 6 R2
19. 8 R2
20. 3 R4
21. 8 R1
22. 4 R2
23. 4 R3
24. 4 R1

Page 16
1. 6 R1
2. 9 R3
3. 8 R1
4. 4 R3
5. 8 R2
6. 5 R1
7. 7 R2
8. 3 R1
9. 7 R4
10. 6 R6
11. 8 R3
12. 5 R4
13. 5 R2
14. 6 R2
15. 5 R1
16. 4 R8
17. 9 R1
18. 8 R6
19. 4 R2
20. 2 R5
21. 4 R4
22. 7 R3
23. 2 R2
24. 2 R1

Page 17
1. 3 R2
2. 8 R3
3. 9 R1
4. 9 R1
5. 7 R2
6. 5 R3
7. 3 R5
8. 5 R4
9. 6 R5
10. 2 R2
11. 3 R4
12. 3 R6
13. 4 R3
14. 3 R1
15. 3 R2
16. 3 R3
17. 7 R5
18. 7 R1
19. 5 R1
20. 3 R7
21. 8 R1
22. 9 R3
23. 8 R1
24. 7 R4

Answer Key ⟩ ◎ ⟩ ◎ ⟩ ◎ ⟩ ◎ ⟩ ◎ ⟩ ◎ ⟩ ◎

Page 18
1. 12 R1
2. 13 R1
3. 16
4. 14 R3
5. 21 R3
6. 23
7. 11 R4
8. 12
9. 12 R3
10. 10 R3
11. 16 R1
12. 12 R1
13. 12 R1
14. 24
15. 18 R1
16. 13 R1
17. 10 R7
18. 29
19. 22 R1
20. 13 R3

Page 19
1. 41
2. 51
3. 71
4. 121
5. 91
6. 61
7. 91
8. 111
9. 81
10. 233
11. 211
12. 111
13. 71
14. 51
15. 213
16. 124
17. 43
18. 61
19. 91
20. 312

Page 20
1. 32
2. 64
3. 56
4. 66
5. 201
6. 105
7. 207
8. 55
9. 105
10. 66
11. 103
12. 102
13. 98
14. 104

15. 74
16. 93
17. 118
18. 108
19. 59
20. 102

Page 21
1. 109
2. 123
3. 209
4. 106
5. 119
6. 46
7. 79
8. 144
9. 133
10. 131
11. 168
12. 114
13. 138
14. 137
15. 140
16. 237
17. 471
18. 143

Page 22
1. 71 R1
2. 87 R1
3. 107 R1
4. 129 R1
5. 116 R1
6. 49 R3
7. 113 R1
8. 258 R3
9. 163 R3
10. 151 R4
11. 149 R2
12. 116 R6
13. 116 R2
14. 134 R6
15. 282 R1
16. 126 R1
17. 223 R1
18. 191 R1

Page 23
1. 118 R3
2. 184 R1
3. 24 R5
4. 242 R8
5. 119 R4
6. 99 R2
7. 244 R1
8. 19 R8
9. 189 R3
10. 192 R1
11. 55 R1
12. 157 R1

13. 234 R1
14. 89 R1
15. 139 R5
16. 52 R3

Page 24
1. 612
2. 509
3. 301
4. 802
5. 301
6. 507
7. 2256
8. 501
9. 721
10. 601
11. 678
12. 811
13. 1001
14. 902
15. 1000

Page 25
1. 448 R1
2. 745 R1
3. 486 R6
4. 407 R1
5. 701 R1
6. 661 R1
7. 903 R1
8. 521 R1
9. 700 R8
10. 260 R5
11. 1337 R2
12. 293 R3
13. 288 R3
14. 445 R2
15. 458 R5
16. 2221 R3

Page 26
1. 81
2. 104
3. 10
4. 43
5. 72
6. 14
7. 136
8. 22
9. 29
10. 124
11. 44
12. 99
13. 478
14. 331
15. 433
16. 321

Page 27
1. 56
2. 64
3. 106
4. 312
5. 78
6. 162
7. 234
8. 81
9. 151
10. 2037
11. 1116
12. 1808
13. 2331
14. 636
15. 4999
16. 578

Page 28
1. 51
2. 26
3. 81
4. 41
5. 48
6. 61
7. 58
8. 46
9. 39
10. 258
11. 137
12. 797
13. 506
14. 709
15. 136
16. 1111

Page 29
1. 41
2. 78
3. 208
4. 137
5. 75
6. 294
7. 174
8. 48
9. 184
10. 741
11. 2730
12. 1334
13. 614
14. 2406
15. 3060
16. 1110

Page 30
1. 42
2. 66
3. 21
4. 28
5. 95
6. 94
7. 9
8. 7
9. 8
10. 766
11. 434
12. 775
13. 87
14. 66
15. 43
16. 56

Page 31
1. 5
2. 16
3. 9
4. 17
5. 35
6. 18
7. 10
8. 6
9. 7
10. 28
11. 34
12. 45
13. 89
14. 390
15. 271
16. 222

Page 32
1. $0.57
2. $0.31
3. $1.01
4. $1.05
5. $0.20
6. $1.02
7. $2.20
8. $0.20
9. $0.71
10. $1.85
11. $0.50
12. $3.59
13. $1.05
14. $0.61
15. $2.49
16. $1.04

Page 33
1. $5.05
2. $6.22
3. $22.11

4. $6.03
5. $1.80
6. $3.16
7. $14.14
8. $5.95
9. $4.43
10. $5.21
11. $5.90
12. $9.80
13. $19.69
14. $23.70
15. $35.99
16. $5.71

Page 34
1. 6
2. 11
3. 17
4. 9
5. 6
6. 12
7. 24
8. 11
9. 48
10. 11
11. 22
12. 5
13. 38
14. 14
15. 15
16. 26

Page 35
1. 8
2. 7
3. 6
4. 14
5. 4
6. 7
7. 24
8. 19
9. 5
10. 16
11. 20
12. 15
13. 9
14. 22
15. 33
16. 13

Page 36
1. 6 R10
2. 4 R20
3. 3 R30
4. 7 R10
5. 3 R30
6. 12 R10
7. 8 R10
8. 11 R10
9. 6 R10
10. 2 R40
11. 4 R20
12. 8 R30
13. 10 R10
14. 8 R10
15. 9 R10
16. 3 R10

Page 37
1. 14 R10
2. 12 R10
3. 14 R10
4. 19 R20
5. 7 R30

6. 48 R10
7. 17 R10
8. 20 R30
9. 13 R40
10. 31 R10
11. 8 R40
12. 22 R30
13. 10 R10
14. 14 R30
15. 49 R10
16. 10 R35

Page 38
1. 31
2. 16
3. 25 candy rolls
4. 26
5. 163 pennies
6. 59
7. 94
8. 14 strips
9. 333
10. 72
11. 8 lollipops, 5 left over
12. 38

Page 39
1. 39 cards
2. 13 cards
3. 32
4. 174, remainder 2
5. 32 seeds
6. 39
7. 36
8. 141
9. 44 kernels
10. 14 pieces
11. 289
12. 96 R2

Page 40
1. B
2. C
3. C
4. D
5. B
6. B
7. C
8. A
9. A
10. A
11. C
12. B
13. A
14. B

Page 41
1. D
2. B
3. A
4. D
5. D
6. A
7. D
8. C
9. A
10. B
11. B
12. C
13. B
14. B

Page 42
1. B
2. C
3. C
4. D
5. B
6. D
7. D
8. A
9. C
10. B
11. D
12. B
13. D
14. A

Page 43
1. C
2. A
3. D
4. D
5. A
6. D
7. D
8. A
9. D
10. B
11. D
12. C
13. D
14. B

Page 44
1. B
2. B
3. A
4. B
5. B
6. B
7. D
8. C
9. A
10. D
11. D
12. C
13. D
14. D

Page 45
1. B
2. C
3. D
4. C
5. D
6. B
7. C
8. A
9. D
10. C
11. D
12. B
13. D
14. C

#3323 Practice Makes Perfect: Division